ORIGIN OF THE QUARTZ PEBBLES

OF THE

SANDSTONE CONGLOMERATE,

AND THE

FORMATION OF THE STRATIFIED SAND ROCKS.

BY

PROF. JEHU BRAINERD

OF CLEVELAND.

READ BEFORE THE AMERICAN ASSOCIATION FOR THE ADVANCEMENT OF SCIENCE,

AT THE CLEVELAND MEETING, JUNE 28, 1853.

CLEVELAND:

PRINTED BY HARRIS, FAIRBANKS, COBB & CO., HERALD OFFICE.

1854.

EXPLANATION.

The following resolution, which was received after this paper was in type, for the minutes of the Cleveland meeting, will explain why the document appears in its present form :—

"*Resolved*, That Prof. Kirtland be instructed to withdraw the paper of Mr. Brainerd from the published proceedings of the Cleveland meeting of the American Association for the Advancement of Science.

"The above is a copy of the resolution as passed by the Association at Washington, 1854. "J. LAWRENCE SMITH,
"Secretary of the Association."

I had until recently supposed that the Association was instituted for the *advancement* of science, and that in accordance with this view, any communication based upon actual discovery, might be admitted.

In regard to the exclusion of this article from the minutes, I have no reflections to cast upon any one. The subject was presented in good faith, and not without a long and patient investigation. If it involves an error, that error should have been candidly met, instead of a suppression of the article. The paper was read before the Association, examined by the committee for publication, and orderd to be printed in the minutes, without solicitation on my part. I have appended the remarks which followed, as reported for the Cleveland Herald, and published at the time. I wish here especially to acknowledge the candor with which Profs. St. John and Kirtland, Col. Whittlesey, Dr. Newberry, and other personal friends, have received and examined the subject; and had the same courtesy been extended by others, there would have been no occasion for this intrusion upon the notice of the scientific world. J. BRAINERD.

ORIGIN OF THE QUARTZ PEBBLES

OF THE

SANDSTONE CONGLOMERATE, ETC.

At the Cincinnati Meeting of the American Association for the Advancement of Science, in May, 1851, the following named subject was presented for examination: "*On Quartz Pebbles of the Sandstone Conglomerate, and reasons for rejecting the Theory of Water Detrition.*" The subject was examined very briefly in a paper read before the Association, and a few pebbles, which contained upon their surface, the impressions of fossil plants, were presented for examination.

The views advanced on that occasion were controverted at the Albany meeting, in August following, by Dr. J. S. Newberry, of this city (Cleveland). His paper was withheld from publication in the Minutes of the Association, but subsequently appeared in the (Hudson) *Family Visiter*. It consisted principally in a reiteration of the old theory of detrition; containing, however, some points purporting to be an explanation of the phenomena presented for consideration — which claims our attention at the present time; there having been no suitable opportunity since the Albany meeting, to bring this subject in its proper light before the members of this body, and the scientific world in general.

Notwithstanding the high degree of esteem in which I hold the character of Dr. Newberry, and his extensive observance of geological phenomena, I cannot admit that his course of reasoning is correct, and especially so in regard to his statement, that the impressions of fossil plants found upon the surface of quartz pebbles, were formed by chemical action subsequent to the deposition of the strata in which these pebbles are found imbedded.

In renewing this investigation, I wish it understood that I am not actuated by a desire to engage in an individual controversy upon "a question of local interest." The positions I have taken, involve in their consideration, the whole subject of the formation of the stratified rocks; and the strong array of facts and arguments which I am now able to present, will, I trust, command from the members of this body an attention and action corresponding to the importance of the question under consideration. It is not my desire to press this subject upon the members of this convention for the purpose of bringing into notice what some may call a vague hypothesis. The whole matter rests upon the observance of facts, and if the positions here taken can be sustained, the prevailing views in regard to the formation of the stratified sand rocks, upon which there seems to be so much harmony of opinion, must be abandoned, and especially so in regard to the formation of quartz pebbles.

I do not expect, gentlemen, that the members of this body, or others who have become interested in the examination of geological phenomena, will adopt this new theory without a critical examination; neither do I desire that this should be the case. It is not an enviable position for any man, to array himself against time-honored views upon scientific subjects; but the discovery and arrangement of facts in science ever has, and ever will, command the attention and respect of the truly learned.

The popular doctrine in relation to the formation of the siliceous stratified rocks is this: that the grains of sand entering into their formation, and constituting their mass, are the grains of silex originally formed in the granite rock, but which have resisted the action of the decomposing agency, and have subsequently been distributed by waves and currents of water, and arranged in strata as we now find them; and that the quartz pebbles were derived from the injected quartz veins found in the granite, broken into fragments, and by their superior hardness resisted the powerful reducing action to which they were subjected, and were left in the form of water-worn or rounded pebbles as we now find them.

In the examination of various treatise upon geology, in relation to this subject, I am surprised at the meagerness of the evidence presented in favor of this hypothesis, which appears to me to be not only not supported by reasonable evidence, but a mere assumption, altogether false in its fundamental principles.

In the absence of all theories it would by comparatively an easy task to present an arrangement of facts, in a systematic order, but in this case, where the ground has been preöccupied by false positions, it becomes first necessary to clear away the rubbish, before we can begin to build the superstructure.

Now there are several insurmountable objections to the theory we have just noticed. First, in regard to the pebbles. There is not the slightest evidence of there having existed at that early period in the formation of the earth's crust, any adequate transporting power. It is, however, admitted that the disintegration of the granite rock, at any geological period, would have formed, by the deposition of its undecomposed grains, a stratified rock like the gneiss, or at least very similar to it, but necessarily confined to the immediate vicinity of the granite, from whence it would have been derived: but in this case, the new formation would contain other minerals found in the granite, besides the grains of silex, for it cannot be supposed that this substance (silex) would alone remain undecomposed. A rock of this character would, therefore, embrace in its composition every variety of igneous rock, both in the form of minute grains and larger masses. These should, however, not possess a uniform figure like the pebbles of the conglomerate, but present every variety of form, from the extreme angular to the smooth rounded pebble. Such a deposition would form a conglomerate entirely unlike that of the sandstone series, and is generally known under the name of "Toadstone," specimens of which are herewith presented for examination.

Again, admitting all that is claimed by the advocates of the old theory, that there has existed an adequate transporting power, and that the quartz pebbles had their origin in the quartz veins of the granite, there are still insurmountable objections that need a better explanation than has yet been given.

It has never yet been shown by what means these pebbles have been arranged in such continuous beds over so vast an area, with scarcely an admixture of the finer grains of sand.

They frequently occur in beds of not more than an inch or two in thickness, extending for hundreds of miles, forming partings to massive beds of rock composed of siliceous grains, almost or entirely destitute of pebbles, and exhibiting no marks whatever of disturbance by waves or running water. The appearance of these strata and intervening bands of quartz pebbles, is represented by the accompanying figure.

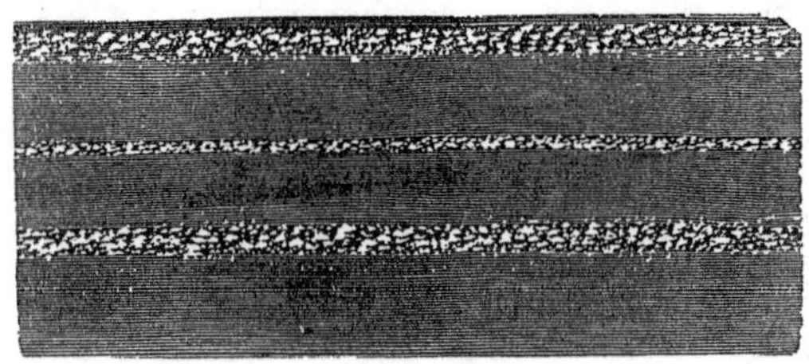

[Fig. 1.]

Another objection which may be urged against the old theory is found in the structure of the pebbles, as compared with the quartz veins from the granite. This difference can be better understood by a comparison of specimens herewith presented, than by any verbal description that can be given. The rounded quartz pebbles possess more the structure of the flint nodules of the chalk; whereas, in the quartz vein, the crystaline structure of the granite is exhibited. So striking is the difference between the two that the most superficial observance is sufficient to detect the variations in the structure.

Another strong argument against the theory of water detrition is brought to view in the fact, that these massive beds of sandstone have been formed in very deep and tranquil water. The almost entire absence of animal remains in these beds shows one of two things, either that the depth of water and consequent distance from shore, was such as to exclude animal life; or that the physical conditions of the waters was such as to be unfavorable to its development, perhaps a combination of both these conditions.

In many places, extending for a distance of more than one hundred and and twenty miles along the out-crop of this formation in Ohio, I have observed the massive beds of sand rock, with the included quartz pebbles, lying in immediate contact with and above the shale, without the least admixture of material, the line of separation being distinct and well-defined. [See Fig. 5.]

Now, if this deposition had been the result of the action of water, in a state of disturbance sufficient to have moved and arranged the pebbles, a mixture of the argillaceous matter of the shale must have taken place. But there are other facts connected with this phenomenon, of a most unequivocal character, which stands opposed to the theory of water detrition, to which reference will again be made.

In agreement with the views here entertained, it may be observed that wherever there is an evidence of the disturbance or arrangement of the material of which the rock is composed, by the action of running water or waves, there is not only an absence of the quartz pebbles, but generally an abundant admixture of argilaceous matter with the silicious grains. The series in such cases is arranged in thin beds of various thicknesses, with partings of clay, containing upon their surface ripple marks, imppressions of rain drops, and enclosing various marine and other fossils, and so far as mere arrangement of material is concerned is wholly mechanical.

Having thus considered some of the objections that may be urged against the old theory, I propose to state definitely the positions I have taken upon this subject, and then introduce a few arguments and facts in their support.

I assume, then, first, that the sand rocks of the Devonian, and probably most, if not all the other systems, have been formed from a *solution* of silicious matter in the waters of the primitive ocean; and second, that the grains of sand and pebbles crystalized from this solution, during their formation in the various geological epochs, have suffered little or no disturbance except upon shoals, the beach of the ocean, or in streams of running water.

It needs not a word of argument to prove that silex can be held in a state of solution in water. Sufficient evidence of this is found in the thermal springs existing in various parts of the globe, especially the Geysers of Iceland; and in the numerous experiments which have been made in the laboratory of the chemist. If further proofs were needed we might refer to the numerous silicates found in mineralogical formations, and to the silicious nodules or flint of the chalk, which have, unequivocally, been formed from solution, and which contain organic remains in their very structure.

We come next to the inquiry: From whence is derived the silex entering into the formation under consideration, and by what means has it been transported—in other words, how have these grains of sand and quartz pebbles originated? There is no controversy among geologists in regard to the antiquity of the granite rock. It is the oldest formation found in the earth's crust, and underlies all the others. It has furnished *material,* in combination with oxygen, carbonic acid, and other gases, for all the stratified series. The principal ingredients entering into the composition of the granite are feldspar, quartz and mica, in variable proportions and variety of texture. A vast amount of this rock was *decomposed* during the formation of the gneiss, which is made up chiefly of that portion which suffered only disintegration.

In this decomposition a vast quantity of feldspar and mica, both of which contain a large per cent. of silex, was brought into a state of *solution,* embracing not only the silex, but also the potash and lime found in its composition, together with many of the metals found also in the granite, and which were subsequently, to a great extent, deposited in metaliferous veins, silicates or ores.

I wish here to notice the difference between matter in a state of mechanical division, held in suspension in the waters of the primitive ocean, and matter in a state of complete solution. In the first instance, the particles in subsiding would arrange themselves according to their specific gravity and the minuteness of division, and at best, would form a mixed and confused mass of matter, entirely unlike the parent rock from which it originated. In the formation of granite a mobility of atoms was allowed, in consequence of the fluidity of the primitive mass by heat, and hence resulted the crystaline character of the formation, and the chemical combination of atoms having an affinity for each other.

We have then in the granite, as the result of this chemical affinity, the two important silicates, feldspar and mica; in the latter, the silicate of alumina, containing 47.95 per cent of silica, and in the former (common feldspar), the silicate of potash, containing 65.21 per cent of silex; and in another variety (albite), the silicate of soda, 69.09 per cent. of silex. In Labradorite, another variety of feldspar, there is 53.42 of silex, and 12.35 of lime.*

A solution is distinguished from a mechanical division, or trituration, by the complete mobility of its atoms, and the readiness with which they obey the law of molecular attraction in the formation of crystals. In this phenomenon we have one of the most sublime operations of nature, without which the whole crust of the earth would have constituted a confused mass, a complete chaos, unfit for the support of either vegetable or animal life.

No geologist would for a moment pretend that the vast beds of lime rock found in the earth's crust, and now forming in the waters of the ocean, have been produced in any other manner than from a solution of the calcareous matter.

There is no source from which this could have been obtained except from the decomposition of the feldspar and mica. We have shown that silex very largely predominates in these minerals, and must also have gone into a state of solution with the calcareous matter with which it was in chemical union.

It is estimated that more than one-half of the bulk of the earth's crust is made up of silica, and in the stratified rocks it very largely predominates over any other mineral.

We ask, then, what other disposition could have been made of this material, than the formation of the sandstone series, from its solution in the primitive waters of the ocean? As nearly as we are able to judge it appears that the proportions of the material in the calcareous and silicious rocks, is about the same as is found in the granite in the two minerals, feldspar and mica. It should, however, be observed that the bulk of the lime rocks have been very greatly increased by the combination of carbon with the lime, derived from carbonic acid (oxygen being the natural solvent of carbon), forming the carbonate of lime, whereas in the feldspar and mica it is a *silicate*.

The homogenous character of the various formations is another

*Note. { Feldspar, Si. 65.21, Al. 18.13, Ka. 16.66. Dana, page 326.
Mica, Si. 47,50 Al. 37.30, Fe. 3.20, Mn. 0.90, K. 9.60. Dana, p. 357

evidence that they have been formed from solution. This remark seems to apply most especially to the rocks under consideration, and also to those of the carboniferous epoch. It is generally maintained that the coal is of vegetable origin, but if the position above taken is correct, it may possibly be shown that even the coal series have been formed from a solution of carbon, first, in the oxygen of the atmosphere, and, second, in the form of carbonic acid, absorbed by water, decomposed and deposited in beds similar to that of the other rocks. Investigations by Charles Whittlesey, Esq., and Daniel Vaughan, strongly point to this origin. One thing may be relied upon as certain, that the coal deposits are not made up of transported material.

I now propose to present a few facts in favor of the position taken, that not only the quartz pebbles, but also the great mass of the sand rocks, have been formed from a solution of silex in the waters of the ocean. A microscopic comparison of the grains of sand from the massive beds alluded to, and those from the beach of the ocean, shows that there is a sufficient amount of difference in the external figures of the two to warrant the conclusion that those from the sand rock have never suffered abrasion from the action of the waves. In those derived form the rock, a great number present well formed crystals, having perfect angles, whereas those from the beach show distinctly that they have suffered detrition.

Nor is this all the difference presented to the eye of the observer. In the former case, the grains appear to be *all* composed of silex, without any mixture of other material, while in the latter case there is every variety of mineral promiscuously mixed together. This condition of things becomes still more apparent when we examine a gravel beach, even in the vicinity of the granite, where all the pebbles are derived from the primitive rock.

We here find hornblend, trap, veinstone, jasper, carnelian, agate, and many other kinds, with every variety of granite, presenting forms from the extreme angular to the nearly spherical, all commingled together ; collected into beds or nests of variable thickness and extent, interspersed with more extended beach of mixed sand of the character above described.

In the sandstone conglomerate the quartz pebbles are arranged in thin horizontal strata of great extent, and so far as my observations have extended, these deposits are entirely free from any form of primitive igneous rocks or other pebbles.

In some situations the sand rock presents thin beds or laminæ, bearing the evident marks of running water upon their surface. In such cases it appears evident that the material, after its crystalization and deposition, has undergone a new arrangement, which is simply due to the mechanical action of the water. These beds generally contain debris from a neighboring shore, fragments of marine or terrestrial plants, shells and fishes.

Another evidence in favor of the formation of the sand rock from a solution is found in the comparison, under the microscope, of the grains of silex from recently disintegrated granite, where they have suffered no detrition, and the grains of silex from the massive beds before alluded to. The silex from the granite presents no well-formed crystals, the grains being irregularly angular, as can be seen with a pocket lens in a specimen of granite. In the sand rock, the crystals are well formed, and characteristic of the mineral.

In stratified rocks, where the *arrangement* of the material is evidently due to the action of water, lime sometimes forms the cementing property, in which case we have a calcareous sand rock.

In the massive beds under consideration there is generally no cementing property except the silex, showing that to a certain extent the silex was in a state of solution.

The structure and external form of the pebbles will next claim our attention.

In size, the quartz pebbles possess a considerable degree of uniformity, ranging from half an inch to an inch or more in diameter, and usually having a flattened or reniform figure. The color is generally of a milky white. They are sometimes nearly transparent, and are often stained throughout with the oxide of iron, and present the color of jaspery quartz. Occasionally a pebble is found of a dark brown color, but they all possess the same general structure and appeaaance.

Their surface is not smooth, as in the *water rolled pebble,* but full of indentations or pits, which can be seen very distinctly with the naked eye. With a lens of moderate power, they become very well defined, showing, conclusively, that when formed they were sufficiently soft to receive indentations from the grains of sand that surrounded them.

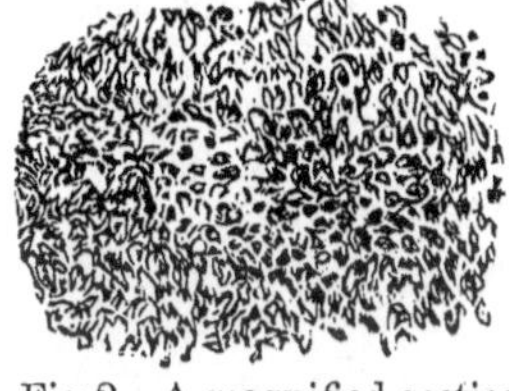

Fig. 2—A magnified section of the surface of a pebble.

Their internal structure is compact and glassy in its appearance, like the flint nodules of the chalk, which are known to have been formed from solution, containing, as they do, the perfect forms of marine infusoria, fragments of sponge, scales of fishes, and other organic remains.

The larger number of these pebbles exhibit solid fracture, though they frequently contain cavities filled with the oxide of iron, but more frequently possess the perfect character of a geode, presenting well defined and beautiful crystals, specimens of which are herewith presented for examination. In one instance I found a tolerably well formed crystal of quartz, projecting from the side of a pebble, which has certainly suffered no detrition in water, to which the attention of the incredulous is invited [See Fig. 3—a]. Not only do the pebbles contain upon their surface the small pits before alluded to, but they also contain larger indentations [b—Fig. 3] made by other pebbles, and which

[Fig. 3.]

exhibit the most unequivocal evidence of their having been, at the time of contact, in a soft and yielding state.

That these pebbles have been formed from a solution — were deposited in the form of a soft gelatinous mass, capable of taking the form or mould of any harder substance upon which they might chance to rest — that they have since suffered no disturbance whatever, or detrition in water — is most conclusively shown by an examination of the specimens herewith introduced, which contain upon their surface the impressions of plants, shells, and other substances.

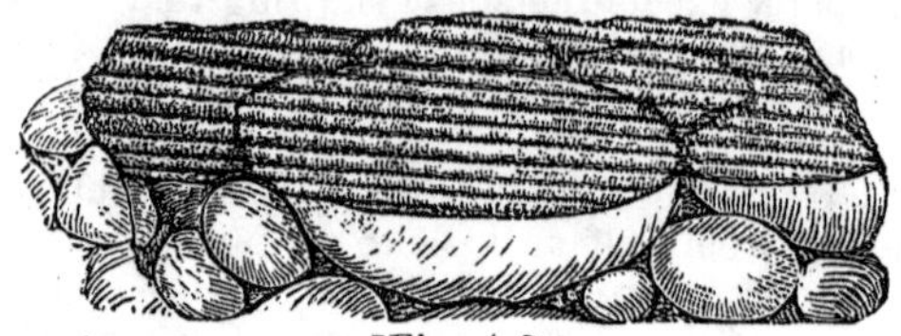

[Fig. 4.]

Fig. 4 shows the impression of the calamite upon the surface of the pebbles.

If further evidence were needed to prove the position we have taken, namely, that the silicious stratified rocks have been formed from solution an abundance of it is at hand. Erman, in his Travels in Siberia, found upon the western slope of the Ural Mountains, compact beds of quartz rock in horizontal strata, containing impressions of corals, star-fishes, species of encrenites, and sea urchins. (Vol. 1, pp. 72—85. Lea & Blanchard, Philadelphia. 1850.) It may be further stated that, according to the observations of this writer, the silicious rock was observed in various places in the district above alluded to, *in the various forms of common granulated sand, that containing quartz pebbles, and the compact fossiliferous stratum above mentioned.*

Compact quartz rock, of much later geological date, generally set down as belonging to the tertiary, is presented for examination in the form of buhr stone, containing numerous fossils.

On what is called Flint Ridge, in Ohio, there is a bed of compact quartz, with cavities beautifully studded with crystals. This stratum belongs to the carboniferous formation. It is generally known as the buhrstone of Ohio, and is used to some extent in the manufacture of millstones. Like the buhr stone of France, this has been formed from solution, which fact is most satisfactorily shown by an examination of the specimens.

Again, the form, structure and arrangement of the flint nodules in the chalk formation, affords still further proof of the correctness of the positions I have taken.

Although it is said by Mr. Leyell, that the flint pebbles upon the beach, at the base of the chalk cliffs, have been rounded and worn to their present form by attrition in water, we find upon examination that these nodules possess the same general form in the strata which have never been disturbed since their formation. Nor is this all; these *undisturbed* nodules of flint exhibit upon their surface indentations produced by the surrounding mass of chalk, in a manner exactly similar to that found on the quartz pebbles of the sandstone conglomerate.

These pebbles or nodules, as above stated, contain organic remains of a very frail and delicate character, showing that the silicious matter was, at the time they were enclosed within it, in a perfectly fluid condition. In confirmation of this I will state that it is affirmed by some writer (Mr. Leyell, I believe), that these flint nodules, and even the quartz pebbles of the older rocks, when first taken from deep mines, are so soft that they can be cut with a knife, or even scratched with the finger nail, without difficulty. One thing is certainly known to every man who has ever been engaged in the sandstone quarries, that the stone, when first raised from their beds, are soft and much more easily worked than after they have been for a time exposed to the action of the atmosphere and, as the miners term it, loose their sap and become seasoned, after which they cannot be softened by immersion in water. The same is true of silex brought to a state of solution in the hands of the chemist. If it is suffered to become dried, it returns again to the solid form of the mineral.

As another instance of the formation of solid strata of silex, may be mentioned the flint mountains of Tennessee. Here in the carboniferous formation, strata of compact flint is found quite extensively developed, specimens of which are presented for examination.

There are one or two more points in relation to the subject under consideration that are worthy of observation.

It is well known that the calamite possessed a hollow jointed stalk like the equisetum of the present day, but having a much larger growth. These are not unfrequently found in the sand rock completely silicified, or having the same internal structure as is found in the surrounding stratum, and often containing quartz pebbles *within* the unfractured stalk. I know of no reasonable explanation of this phenomenon except the one advocated in this paper.

The arrangement of the quartz pebbles in thin horizontal beds of great extent, shows a uniformity of action in the depositing agent, not easily explained upon the hypothesis of water detrition, but which can be readily accounted for upon the principle of chemical action or deposition from a solution.

I have one feature more to present for examination, accompanied with specimens and drawings which in itself alone, independent of all other facts brought to view, is sufficient to set aside the theory of water detrition, and to establish the position we advocate.

In an extended district in the northern part of this state (Ohio), the conglomerate rests directly upon a stratum of soft, dark colored, argilaceous shale, evidently deposited in still deep water, and finely laminated, showing no evidence whatever of disturbance by the action of waves or running water.

A thin band of white quartz pebbles lies in immediate contact with and above this formation, without the least admixture of the strata. I have traced this bed for more than one hundred miles, and find this feature perfectly uniform. But what is still more unaccountable upon the old hypothesis, is the fact that the pebbles are flattened upon their under surface, exhibiting unequivocal marks of their having been deposited upon the smooth surface of the shale in a soft or gelatinous state, where they remained in a perfectly quiet state until induration had taken place, and from which position they have not since been disturbed, as seen in the annexed figure.

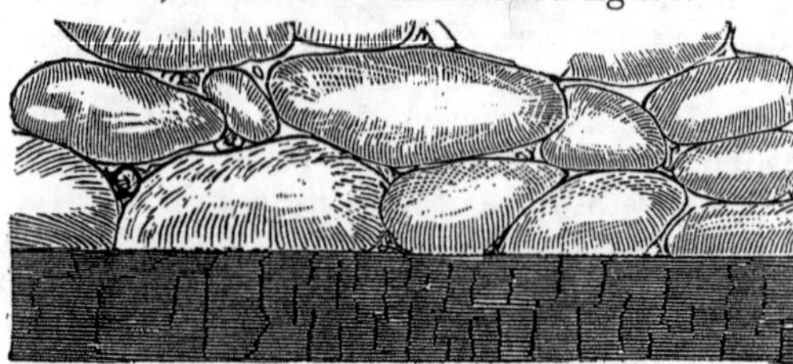

[Fig. 5.]

In conclusion, permit me to say that here are FACTS from the records of Nature that *cannot* be controverted.

They are not to be looked upon as uncertain witnesses, bearing false testimony.

They are not isolated specimens which can be explained away as the mere freaks of Nature.

The evidence introduced by them is a quotation from the pages of the natural history of the globe, written in imperishable characters upon the face of the flinty rock.

There it stands: the chapter and page, and even the date of the volume is given.

Gentlemen of this Association, in view of all these facts, what is your decision upon this important question in Geology? Shall we, without evidence, adhere to the old theories, or shall we cast them aside as unworthy of our confidence, and adopt the view here presented, which appears to be supported by such unimpeachable evidence?

The following report of remarks is taken from the Cleveland Herald, of July 30:

Dr. Newberry remarked in reference to the paper of Professor Brainerd, as follows: That this discussion arose from the discovery by Prof. B. of specimens from the conglomerate, in which the impressions of plants are as distinctly transmitted to the pebbles which are scattered through the rock, as to the interspaces of indurated sand. Prof. B. inferred from these specimens, that the quartz pebbles had been formed syn-chronous with, or anterior to the deposition of the rock. Dr. N. said if this theory were accepted, it would entirely revolutionize our notion of the mode of formation of not only this important rock, but of a large part of the silicious portion of the Earth's crust. Before abandoning the time-hallowed opinion, that these pebbles were fragments of quartz rock rounded by attrition, he would look for some cause which could have eroded the pebbles subsequent to their deposition. Quartz is nearly pure silicia—silicia is so readily soluble, by water—alkaline solutions, &c., has so wide a range of chemical combinations, that all the phenomena of these flattened and eroded pebbles might easily be produced by chemical action. Dr. N. stated that he had also found mingled with the quartz pebbles of the conglomerate, rolled fragments of jasper, trap, silicious stratified slates, &c., pebbles which he recognized as fragments of well known rocks.

The structure of the quartz pebbles was exactly that of the massive quartz, and not that of geodes, septaria, chalk-flints &c., and without central cavities, or a concentric structure. He said he had recently seen on the south shore of Lake Superior, a conglomerate in the process of formation, entirely by mechanical means, which could in no respect be distinguished from the conglomerate of the coal measures. The instance cited was where the Potsdam Sandstone, trap, quartz, jasper &c. are broken down together, forming sand and rounded pebbles, which are in form, mineralogical character, and in every respect the same with the pebbles under consideration.

In regard to the presence of quartz pebbles in the interior of certain fossil plants, he said that such plants are hollow and reed-like, and he had frequently found well preserved fossil nuts in the stems of the same plants.

In reference to the theory of Prof. B. that the coal was of mineral origin, he would say that there seemed to be no room for a doubt, that all coal was of vegetable origin; because, first, beds of coal are associated with great numbers of fossil plants, each plant being covered with a coating of carbonaceous matter, corresponding in thickness to the size and density of the original plant; second, every variety of coal, whether anthracite, cannel or bituminous, when properly treated and examined with the microscope, was found to be composed almost entirely of the remains of vegetable tissue. Coal had been formed from wood by artificial processes, of which he gave some new and interesting examples.

Remarks of Prof. Hall, of Albany, N. Y., on the paper read by Prof. Brainerd, of Cleveland.—*Herald.*

Prof. Hall said: That the paper just read struck at all former views on the subject of the formation of the stratified rocks, and he did not feel disposed to spend his time, and that of the members, in discussing a fact which was considered so well established. Prof. B., he said, had claimed the whole ground, and if his views were adopted, it would essentially modify all former teachings upon the subject.

The only object he had in view in offering his remarks in the negative, was this, that the paper read by Mr. Brainerd would go into the minutes of the Association, and if nothing was said to the contrary, it must be inferred that the premises claimed by him are correct; this he could not consent to; he adhered to the old doctrine, that these pebbles were water-worn fragments of granite quartz veins, and the grains of sand constituting the mass of the sand rock, were also derived from the granite. Mr. Brainerd had stated that the granite was the oldest rock found in the crust of the Earth, and forms a nucleus upon which all the stratified rocks rest. This he denied. He stated that all the granites of New England had been made up from the destruction of the earlier stratified rocks—including those of the fossiliferous period. He did not think that the subject was of sufficient importance to occupy *his* time or that of the Association; but he would say in reference to the specimens presented for examination, that the impressions had been undoubtedly formed by chemical action. He admitted that the flint nodules—the burr stone—the flint ridge of Ohio—and the beds of horn stone from the Tennessee mountains; and also, rock crystals,—had been formed from solution of silex, in the manner claimed by Mr. Brainerd, but he could not admit that the rounded quarts pebbles and silicious grains constituting the mass of the silicious stratified rocks, had been formed in this way. He was willing to yield half the ground to Mr. Brainerd, and thought he ought to be satisfied with that.

Pending these remarks the Association adjourned to 3 o'clock, P. M.

3 o'clock, P. M.

Professor Brainerd expressed a desire to reply to the remarks of Prof. Hall, and five minutes time was granted for that purpose.

He claimed that Prof. Hall had not touched the merits of the question—that he did not wish to continue the discussion for any purpose except to arrive at the facts in the case.

He considered the evidence he had presented in favor of the new theory, to be overwhelming in its character, and claimed that Prof. Hall had not touched the merits of the case. He claimed that it belonged to those who hold to the attrition theory to show, by what agency the fragments of quartz rock had been transported.

He wished Mr. Hall to explain by what means the pebbles had been flattened upon their under surface by contact with the shale—how indentations have been formed upon the surface of the pebbles, and especially how well defined crystals could be preserved upon the surface of the pebbles during the process of attrition.

He desired to have the *facts* he had presented fully explained, and he thought this could only be done upon the new theory he had presented.

[Here the report for the Herald closes.]

The author of the article in question, feels it his privilege, under the circumstances, to mention a few more facts that were brought to view on that occasion. The attention of Prof. Hall was called to the fact, that in the massive beds of stratified sand rock, the grains of sand present well defined crystal faces,—that under the microscope this feature becomes very apparent, showing conclusively that these grains have not suffered abrasion from the action of waves or running water; he was called upon to show what disposition had been made of the silicious matter that had gone into a state of solution, in the decomposition of the primitive rock during the formation of the gneiss; or to show any analogous case of chemical action, which he claimed had caused the impressions of plants upon the surface of flint or quartz pebbles. He was also called upon to show where he would draw the line between the rocks formed from *chemical* action, and those formed by *mechanical* action. The *facts* presented have not been explained.

www.ingramcontent.com/pod-product-compliance
Lightning Source LLC
LaVergne TN
LVHW020641110826
845149LV00004B/1312

* 9 7 8 1 4 1 8 1 9 0 5 6 9 *